BEAUTIFUL BIOMES

FOREST BIOME

by Elizabeth Andrews

Cody Koala
An Imprint of Pop!
popbooksonline.com

abdobooks.com

Published by Pop!, a division of ABDO, PO Box 398166, Minneapolis, Minnesota 55439. Copyright ©2022 by Abdo Consulting Group, Inc. International copyrights reserved in all countries. No part of this book may be reproduced in any form without written permission from the publisher. Cody Koala™ is a trademark and logo of Pop!.

Printed in the United States of America, North Mankato, Minnesota

102021
012022

THIS BOOK CONTAINS RECYCLED MATERIALS

Cover Photo: Shutterstock Images
Interior Photos: Shutterstock Images, 1, 6, 9, 13, 19; RondaKimbrow/ Getty Images, 5 (top); mihtiander/Getty Images, 5 (bottom left); lucky-photographer/Getty Images, 5 (bottom center); Jillian Cooper/Getty Images, 10–11; THEPALMER/Getty Images, 14; Burachet/Getty Images, 15; Mirosław Nowaczyk/Alamy Stock Photo, 16; pablo_arriaran/Getty Images, 20

Editor: Tyler Gieseke
Series Designer: Laura Graphenteen

Library of Congress Control Number: 2021942250

Publisher's Cataloging-in-Publication Data

Names: Andrews, Elizabeth, author.
Title: Forest biome / by Elizabeth Andrews
Description: Minneapolis, Minnesota : Pop!, 2022 | Series: Beautiful biomes | Includes online resources and index.
Identifiers: ISBN 9781098241018 (lib. bdg.) | ISBN 9781098241711 (ebook)
Subjects: LCSH: Forests and forestry--Juvenile literature. | Biotic communities--Juvenile literature. | Habitats--Juvenile literature. | Life zones--Juvenile literature. | Forest animals--Juvenile literature. | Forest plants--Juvenile literature. | Forest ecology--Juvenile literature.
Classification: DDC 577.3--dc23

Hello! My name is

Cody Koala

Pop open this book and you'll find QR codes like this one, loaded with information, so you can learn even more!

Scan this code* and others like it while you read,

or visit the website below to make this book pop.

popbooksonline.com/forest-biome

*Scanning QR codes requires a web-enabled smart device with a QR code reader app and a camera.

Table of Contents

Chapter 1

Among the Trees

A biome is a large, natural area. It is known for the plants and animals that live there, and its **climate**.

marine

desert

forest

Watch a video here!

A forest biome is an area covered with trees and woody plants. About one-third of all land is covered by forests. There are three kinds of forests.

Chapter 2

Temperate Forests

Temperate forests have mild **climates**. They are not too cold or too hot. Temperate forests experience four seasons. Leaves fall in autumn and come back in spring.

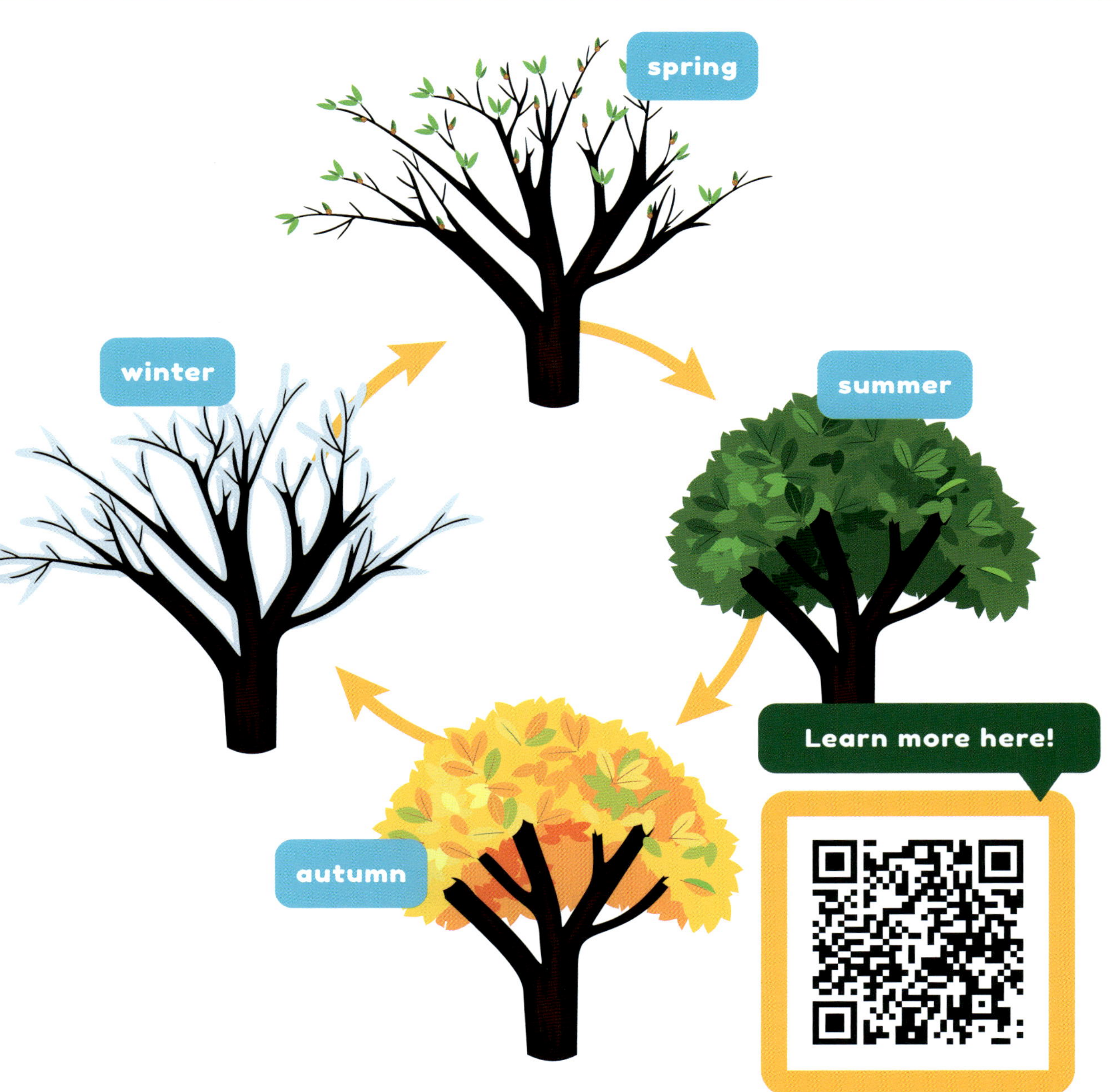

spring
winter
summer
autumn
Learn more here!

Animals like black bears
and deer live among large
maple and oak trees. Ferns
and wildflowers grow on the

forest floor. Insects, birds, and rodents also live there. They make their homes in the trees or on the ground.

Chapter 3

Tropical Forests

Tropical forests grow near the **equator**. Tropical forests can receive up to 400 inches (1,000 cm) of rain each year! The weather is hot and **humid** all year long.

Learn more here!

Half of Earth's plant
and animal **species** live in
tropical forests.

Trees grow large due to rain and warm weather. Many of the animals eat fruit! Meat-eaters like jaguars and anacondas also live there.

The Wasai tree roots are good for kidney health.

Tropical forests are helpful biomes. They are home to plants that are used to make life-saving medicines. The plants in these biomes also take in **greenhouse gases** that hurt the Earth.

Boreal Forests

Boreal forests are found far north of the **equator**. Their winters are long, snowy, and cold. Warmer weather in the summer brings the most **precipitation** for these forests.

Boreal forests are also called taigas.
Complete an activity here!

Evergreens are the most common trees. They do not drop their leaves in winter. **Mammals** that can handle the cold **climate**, like bears and moose, live in boreal forests. Woodpeckers and chickadees make homes in the trees.

Making Connections

Text-to-Self

Which kind of forest would you want to live in? Write a few sentences about why you chose this forest.

Text-to-Text

Have you read a book about any other biomes? If so, how are they similar to and different from forests?

Text-to-World

How do you think forests help the world? Is it important to protect them?

Glossary

climate – weather conditions that are usual in an area over a long period of time.

equator – an imaginary line around the middle of Earth, halfway between the North and South poles.

greenhouse gases – gases, such as carbon dioxide, that trap heat in Earth's atmosphere.

humid – having a high amount of moisture in the air.

mammal – an animal that makes milk to feed its young and usually has hair or fur on its skin.

precipitation – moisture such as rain, hail, or snow that falls to Earth.

species – living things that are very much alike.

Index

popbooksonline.com

Thanks for reading this Cody Koala book!

Scan this code* and others like it in this book, or visit the website below to make this book pop!

popbooksonline.com/forest-biome

*Scanning QR codes requires a web-enabled smart device with a QR code reader app and a camera.